REASONOVER'S

LAND MEASURES

By J. ROY REASONOVER

Revised, with an introduction
by Michelle M. Haas

Copano Bay Press

INTRODUCTION

The rare book market is a fickle beast. Authors and genres saunter in and out of fashion in the collectible realm just as they do in the world of new books. Today's bestseller may well be tomorrow's Bargain Bin feature. But books of merit will always sell, and will sell well decades after the printing presses cease to cough up new copies. Mr. Reasonover's original 1946 treatise on land measures doesn't linger long on the shelves of used bookstores. It is snapped up by history buffs, surveyors and other land professionals. It is for this reason, and as a tribute to Mr. Reasonover's fine work and persistence, that it is being republished.

Creating the first edition was clearly a labor of love for Mr. Reasonover and the design of this second edition provided a similar gratification to me and all who worked on the project. The original format, created without the aid of computers, was a bit cumbersome. The design has been updated to make the information easier to access and process. A few bits of extraneous commentary relating to the commonly known history of Texas have been expunged. But the calculations themselves, performed without the use of electronic calculators or personal computers, are the guts of the book and thus remain unchanged. Mr. Reasonover, armed with his slide rule and ardent diligence, produced a remarkable reference in 1946. Sixty years later the information is still relevant and necessary. This new edition has simply been edited to make the data more fit for convenient consumption.

Michelle M. Haas
Seven Palms - Rockport, Texas

CONTENTS

CONTENTS Continued

PREFACE

When I entered the land department of The Federal Land Bank of Houston in March 1932, I felt the need for various information pertaining to land measures and quantities that was not readily available at the time. During the next nine years I computed and compiled much information on the subject, originally intending to publish it as supplemental material in a specifically arranged book of "Field Traverse Tables". However, the book had not been completed when I was called to active duty with the Army in March 1941, and since that time I haven't had occasion to do any more about it. I have decided, however, after five years, that the information herein should be published as a reference work for the use of abstractors, architects, oil company officials and employees, lawyers, land surveyors, land appraisers, civil engineers, engineering students, law students, realtors and others whose occupation, business or professional duties require reference to facts pertaining to land measures and quantities.

J.R. Reasonover
Lieutenant Colonel
Army of the United States
15 March, 1946

French and Spanish Land Measures in the United States and Canada

A glance into land records in Texas and Louisiana reveals more than merely feet, yards and acres. The provenance of much land in North America includes time spent in the hands of Spain and France. Herein are explanations of a number of older land measures of France, Spain and Mexico still in use today in parts of the United States and Canada, together with their English equivalents and conversion factors.

Old French, Spanish and Mexican Fundamental Units of Length and Area

French Units

1 pouce	= 12 lignes
1 pied	= 12 pouces
1 toise	= 6 pieds
1 perche	= 3 toises
1 perche	= 18 pieds
1 arpent	= 10 perches
1 arpent	= 30 toises
1 arpent	= 180 pieds
1 lieue	= 84 arpents
1 lieue carrée	= 7,056 arpents carrées

Spanish and Mexican Units

1 linea	= 12 puntos
1 pulgada	= 12 lineas
1 pié	= 12 pulgadas
1 vara	= 3 piés
1 cordél	= 50 varas
1 legua	= 5,000 varas
1 caballería	= 609,408 varas cuadradas
1 labór	= 1,000,000 varas cuadradas
1 sitio	= 25,000,000 varas cuadradas

Comparison of English, French and Spanish Leagues in the United States and Canada

The English League

An English league is 3 miles or 16,840 feet in length. A square English league contains 9 square miles or 5,760 acres.

The French League

A lieue, or French league, is 84 arpents in length. A lieue carrée, or square French league, contains 7,056 superficial arpents.

A French league in Missouri and other territory formerly part of the Spanish province of Upper Louisiana is 3.0625 miles or 16,170 feet in length based on the French lineal arpent which is equivalent to 192.5 feet in that region.

A French league in the Canadian province of Quebec is 3.0519 miles or 16,114.14 feet in length based on the French lineal arpent which is equivalent to 191.835 feet in Quebec. A square French league in the same province contains 5,961.099 acres.

The Spanish League

A legua, or Spanish league, is 5,000 varas in length. A sitio, or square Spanish league, contains 25,000,000 square varas.

A Spanish league in Texas is 2.63 miles or 13,888.889 feet in length based on the Spanish vara being equivalent to 33 $1/3$ inches in Texas. A sitio in Texas contains 4,428.4 acres.

English Units—
Linear & Superficial

1 inch	= 1,000 mils
1 foot	= 12 inches
1 yard	= 3 feet
1 rod	= 16 $^{1}/_{2}$ feet
1 rod	= 5 $^{1}/_{2}$ yards
1 pole	= 1 rod
1 perch	= 1 rod
1 furlong	= 40 rods
1 furlong	= 660 feet
1 mile	= 5,280 feet
1 link	= 7.92 inches
1 chain	= 66 feet
1 chain	= 100 links
1 league	= 3 miles
1 military pace	= 2 $^{1}/_{2}$ feet
1 square foot	= 144 square inches
1 square yard	= 9 square feet
1 square rod	= 30 $^{1}/_{4}$ square yards
1 acre	= 208.7103258 feet square
1 acre	= 43,560 square feet
1 acre	= 10 square chains
1 rood	= 40 square rods
1 rood	= $^{1}/_{4}$ acre
1 square chain	= 16 square rods
1 square chain	= $^{1}/_{10}$ acre

Metric Measures

1 millimeter	= 0.001 meters
1 millimeter	= 0.03937 inches
1 centimeter	= 0.01 meters
1 centimeter	= 0.3937 inches
1 decimeter	= 0.1 meters
1 decimeter	= 3.937 inches
1 meter	= 39.37 inches
1 meter	= 1.0936 yards
1 meter	= 3.28083 feet
1 dekameter	= 10 meters
1 dekameter	= 32.8083 feet
1 hectometer	= 100 meters
1 hectometer	= 328.083 feet
1 hectometer	= side of square hectare
1 kilometer	= 1,000 meters
1 kilometer	= 3,280.83 feet
1 kilometer	= 0.62137 miles
1 kilometer	= approx. $\frac{5}{8}$ miles
1 myriameter	= 10,000 meters
1 myriameter	= 6.2137 miles
1 hectare	= 10,000 square meters
1 hectare	= 2.471 acres
1 are	= 100 square meters
1 are	= 0.02471 acres
1 centare	= 1 square meter

Comparison of Acre & Hectare

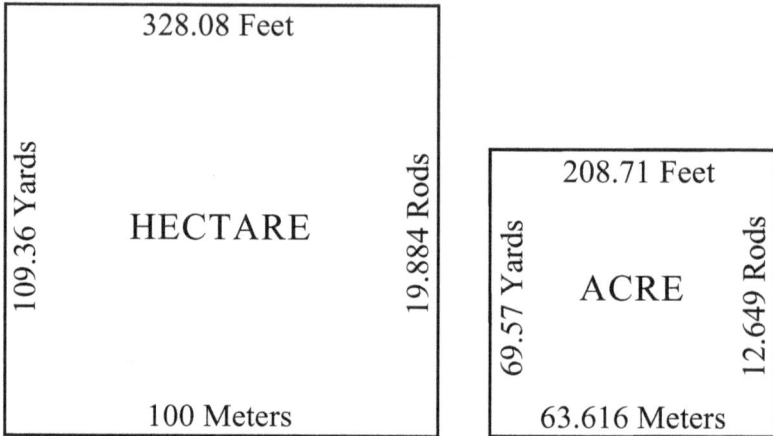

```
                328.08 Feet
  ┌─────────────────────────────────────┐
  │                                      │
1 │                                  1   │
0 │                                  9.  │
9 │           HECTARE                8   │
. │                                  8   │
3 │                                  4   │
6 │                                      │
  │ Y                                R   │
  │ a                                o   │
  │ r                                d   │
  │ d              100 Meters        s   │
  │ s                                    │
  └─────────────────────────────────────┘
            1  H e c t o m e t e r
```

```
               208.71 Feet
        ┌──────────────────────────┐
        │                      1    │
      6 │                      2.    │
      9 │        ACRE          6     │
      . │                      4     │
      5 │                      9     │
      7 │                            │
        │ Y                    R     │
        │ a                    o     │
        │ r     63.616 Meters  d     │
        │ d                    s     │
        │ s                          │
        └──────────────────────────┘
```

1 acre	=	43,560 square feet
1 hectare	=	107,638.7 square feet
1 acre	=	208.71 feet square
1 hectare	=	328.08 feet square
1 acre	=	0.404687 hectares
1 hectare	=	2.471 acres
1 acre	=	63.616 meters square
1 hectare	=	100 meters square
1 acre	=	4,046.87 square meters
1 hectare	=	10,000 square meters

To convert acres to hectares, multiply by 0.404687.
To convert hectares to acres, multiply by 2.471.

Ninety Dimensions of One Acre
Any of the following dimensions include a one acre plot:

10 x 4,356	feet	10 x 404.687	meters	1 x 160	rods
20 x 2,178	feet	20 x 202.344	meters	2 x 80	rods
30 x 1,452	feet	30 x 134.896	meters	3 x 53.34	rods
40 x 1,089	feet	40 x 101.172	meters	4 x 40	rods
50 x 871.2	feet	50 x 80.937	meters	5 x 32	rods
60 x 726	feet	60 x 67.448	meters	6 x 26.67	rods
70 x 622.29	feet	70 x 57.812	meters	7 x 22.86	rods
80 x 544.5	feet	80 x 50.586	meters	8 x 20	rods
90 x 484	feet	90 x 44.965	meters	9 x 17.78	rods
100 x 435.6	feet	100 x 40.469	meters	10 x 16	rods
110 x 396	feet	110 x 36.790	meters	11 x 14.55	rods
120 x 363	feet	120 x 33.724	meters	12 x 13.34	rods
130 x 335.1	feet	130 x 31.130	meters	13 x 12.3	rods
140 x 311.14	feet	140 x 28.906	meters	14 x 11.42	rods
150 x 290.4	feet	150 x 26.979	meters	15 x 10.67	rods
10 x 484	yards	10 x 564.54	varas	1 x 10	chains
20 x 242	yards	20 x 282.27	varas	2 x 5	chains
30 x 161.33	yards	30 x 188.18	varas	3 x 3.33	chains
40 x 121	yards	40 x 141.135	varas	4 x 2.5	chains
50 x 96.8	yards	50 x 112.91	varas	5 x 2	chains
60 x 80.66	yards	60 x 94.09	varas	6 x 1.67	chains
70 x 69.14	yards	70 x 80.648	varas	7 x 1.43	chains
80 x 60.5	yards	80 x 70.567	varas	8 x 1.25	chains
90 x 53.78	yards	90 x 62.723	varas	9 x 1.11	chains
100 x 48.4	yards	100 x 56.454	varas	10 x 1	chains
110 x 44	yards	110 x 51.32	varas	11 x 0.91	chains
120 x 40.33	yards	120 x 47	varas	12 x 0.83	chains
130 x 37.23	yards	130 x 43.42	varas	13 x 0.76	chains
140 x 34.57	yards	140 x 40.32	varas	14 x 0.71	chains
150 x 32.26	yards	150 x 37.63	varas	15 x 0.67	chains

Reduction of Square Units to Acres by Division

When reducing square units to acres by division, the following tables may be found convenient in multiplying the divisor.

Square feet divided by 43,560 equal acres	Square varas divided by 5,645.4 equal acres
43,560 x 2 = 87,120	56,454 x 2 = 112,908
43,560 x 3 = 130,680	56,454 x 3 = 169,362
43,560 x 4 = 174,240	56,454 x 4 = 225,816
43,560 x 5 = 217,800	56,454 x 5 = 282,270
43,560 x 6 = 261,360	56,454 x 6 = 338,724
43,560 x 7 = 304,920	56,454 x 7 = 395,178
43,560 x 8 = 348,480	56,454 x 8 = 451,632
43,560 x 9 = 392,040	56,454 x 9 = 508,086

Square yards divided by 4,840 equal acres	Square varas divided by 5,645.376 equal acres
4,840 x 2 = 9,680	5,645,376 x 2 = 11,290,752
4,840 x 3 = 14,520	5,645,376 x 3 = 16,936,128
4,840 x 4 = 19,360	5,645,376 x 4 = 22,581,504
4,840 x 5 = 24,200	5,645,376 x 5 = 28,226,880
4,840 x 6 = 29,040	5,645,376 x 6 = 33,872,256
4,840 x 7 = 33,880	5,645,376 x 7 = 39,517,632
4,840 x 8 = 38,720	5,645,376 x 8 = 45,163,008
4,840 x 9 = 43,560	5,645,376 x 9 = 50,808,384

Overview of Spanish Measures

The earliest Spanish law establishing standards of land grants to be made in New Spain was the cédula, or decree, of June 18, 1513. Drafted under the King of Spain, Ferdinand V, it provided that land would be granted to settlers in a new territory in amounts of caballerías and peonías, which were composed of fanegas and huebras. These grants were graduated according to the rank of the individual and type of land granted.

A fanega is $1/12$ of a caballería. The term originated from a name applied to a unit of volume used for grain. As a land area measure, the term "fanega" was used in Spanish-speaking lands to describe the amount of land that could be sown with a fanega of seed. The huebra, roughly the Spanish equivalent of an acre, was described as the amount of land a yoke of oxen could plough in one day. There were approximately 9 huebras in a fanega.

The peonía was a grant of land given to foot soldiers, laborers or peónes in a conquered territory. It was defined in the decree of 1513 as consisting of a building lot 50 Castilian feet wide by 100 feet long, 100 fanegas of cultivable land for wheat or barley, 10 fanegas for corn, 2 huebras for a vegetable garden, 8 huebras for planting other trees that grow on dry land, and pasture for 10 brood swine, 20 cows, 5 mares, 100 sheep, and 20 goats.

The caballería was granted to soldiers of the caballería, or cavalry, and was roughly 4 or 5 times as much land as a peonía. It was defined in the decree of 1513 as consisting of a building lot 100 Castilian feet wide by 200 feet long, 500 fanegas of cultivable land for wheat or barley, 50 fanegas for corn, 10 huebras of land for a vegetable garden, 40 huebras for planting other trees that grow on dry land, and pasture for 50 breeding sows, 100 cows, 25 mares, 500 sheep, and 100 goats.

The caballería is now defined as a right-angled parallelogram measuring 524 × 1,104 varas and having a surface equivalent of 105.756 acres in Mexico, 107.948 acres in Texas (108 acres for practical purposes). The other quantities mentioned in the decree of 1513 have no practical significance in the United States today.

Spanish Measures in Texas

In Texas, the Spanish vara is still used in land measurement in addition to feet, rods or poles, yards and chains. The English equivalents derived from the vara and shown herein are based upon the vara's established length of 33$\frac{1}{3}$ inches in Texas. Although the value had been widely used in Texas since the early days of Anglo-American colonization, it was legally set at 33$\frac{1}{3}$ inches by an act of the Texas State Legislature in 1919.

In 1750 the Spanish government made a grant of 50 sitios de gañado menor (50 square Spanish leagues of land), suitable for raising small livestock, on the north side of the Rio Grande River. This grant is considered the first Spanish colonial grant on the north banks of the Rio Grande. The original grantee was Capt. Jose Vasquez Borrego, who was given the land in recognition of meritorious service rendered the Spanish crown in subduing hostile Indians. In 1753 the grant was increased by 25 sitios de gañado mayor (25 square Spanish leagues), suitable for raising large livestock such as cattle.

In 1767, during the reign of King Carlos III of Spain, a specially appointed and empowered Spanish Royal Commission paid a visit to the Rio Grande Valley and instituted a system of colonization in several jurisdictions along the river. The Spanish chain, the official standard, used in surveying land was the cordél, measuring 50 varas in length or 138.889 feet. Land fronting the river was surveyed and granted in long narrow strips of various lengths and widths called porciónes. Due to the meanders along the river, the two longer sides of a riparial porción were rarely the same length. Thus these porciónes consisted of varying amounts of land. The system was similar to that employed by the French in locating land grants on water courses in Michigan, Wisconsin, Canada, etc and by both the French and Spanish in provincial Louisiana. The majority of porciónes with frontage on the Rio Grande River were assigned to the original grantees prior to 1800.

The 1820's were turbulent times in Mexico, to say the least. Within the first 3 years of the decade, the country won independence from Spain, saw its first emperor crowned and dethroned, and ultimately adopted a federal system of government. During his brief reign, Emperor Augustin I decreed Mexico's first colonization law known as the Imperial Colonization Act of 1823.

Provisions in Articles 5 and 7 of the Act established the vara as 3 geometrical pié, referring to the Castilian pié or Spanish foot, which is the equivalent to approximately $9/10$ of an English foot. The pié consisted of 12 Spanish pulgadas. Neither the pié nor the pulgada are used in contemporary Texas land measures, as the vara is used decimally instead. The Articles further established that a straight line of 5,000 varas comprised a league. In modern use, the league in Texas refers to a block of land containing 4,428.4 acres (25,000,000 square varas or 25 labóres). A square, each side of which was one legua or league, was to be called a sitio. 5 sitios composed 1 hacienda. One labór was composed of 1,000,000 square varas (1,000,000 varas on each side, or 177.136 acres). According to the Act, labóres were only to be divided into halves and quarters.

As the year 1824 dawned, Mexico was no longer under the reign of Augustin I and new colonization legislation was on the horizon, as was the empresario system. Much of the land in the present state of Texas lay in the Mexican state of Coahuila, while a portion between the Nueces and Rio Grande Rivers was part of the state of Tamaulipas and a strip of land west of the Pecos River was a part of Chihuahua. The National Colonization Law of August 1824 granted that the Mexican states were authorized to form their own colonization laws provided that they conformed with the articles laid out in the national law. Under the various state laws the vara, labór and sitio remained unchanged from their previous incarnations.

Article 4 of the National Colonization Law prohibited the location of land grants within 20 leagues (52.6 miles) of the United States border and 10 leagues (26.3 miles) from the coast of the Gulf of Mexico without special permission from the Mexican government.

Article 12 provided that no more than one square league of land suitable for irrigation, 4 square leagues of arable land without irrigation facilities, and 6 square leagues of grazing land could be united in the same hands with right of property. This marked the origin of the "11 League Grants" containing 275,000,000 square varas of land.

On March 24, 1825, the Colonization Law of the State of Coahuila and Texas was adopted. By Article 11, the vara of three geometrical piés was adopted as the linear unit for measuring land. The superficial measures adopted were the sitio and the labór, each having the same dimensions as set forth in the Imperial Colonization Act of 1823. Article 34 of the 1825 law outlined an ayuntamiento, or township, as consisting of 4 square leagues, the area of which might be rectangular or irregular in form as was agreeable to the situation.

The Colonization Law of the State of Coahuila and Texas of 1825 was then supplanted by the Law of April 28, 1832. The land measures were retained but the prohibition of land sales near the gulf coast was lifted. Under the new law, land within 10 littoral leagues of the coast of the Gulf of Mexico could be sold at various fixed prices per sitio for the different classes of land.

Under a new Coahuila and Texas state decree of March 1834, the vara was still defined as consisting of 3 geometrical piés and a millonada consisted of 1,000,000 square varas, or a square measuring 1,000 varas on each side. The millonada was, therefore, equivalent to the labór used under previous legislation. Under the new law, the same person was not permitted to purchase more than 275 millonadas (11 square leagues), which was in keeping with the National Colonization Law of 1824 prohibiting more than 11 square leagues to be united in the same hands.

On March 17, 1836, the vara entered into the laws of the new Republic of Texas by a constitutional provision which provided that all laws then in force and not inconsistent with the constitution should remain "in full force until declared void, repealed, altered or expire by their own limitation."

By the provisions of Article 4 of the Secret Agreement in the Treaty of Velasco, and to a less explicit extent in the Public Agreement, Texas claimed the Rio Grande River as its southern boundary. An act of the Texas Congress of December 19, 1836 declared that the civil and political jurisdiction of the Republic extended from the mouth of the Sabine River, along the Gulf of Mexico three leagues from land, to the mouth of the Rio Grande River, thence up the principal stream of that river to its source, thence due north tothe 42nd degree of north latitude, thence along the boundary line as defined by the Treaty of Amity, Settlement and Limits between the United States and Spain, to the point of beginning.

Some parts of west Texas have been sectioned similarly in principle to the rectangular system of surveying public land states. However, in addition to townships and ranges there are sections, blocks, leagues, labóres and surveys of original grantees. A regular section is 1,900.8 varas square (1 square mile) and contains 640 acres. Some fractional sections contain less than 640 acres and some elongated sections contain more. Some land not divided into sections or surveys is composed of leagues usually 5,000 varas square. Some of the leagues in this region of Texas are divided into 25 labóres containing 177.136 acres each.

The vara was also used in the province of Alta California during Spanish and Mexican rule where its length was 32.9927 inches, or 33 inches for practical purposes. By Article 20 of the Mexican Ordinance for Land and Sea of September 15, 1837, the sitio de gañado mayor, sometimes referred to as a "California League", was 5,000 varas square. It contained 4,338.464 acres in the state of California, giving the vara a length of 32.99311 inches according to that ordinance. After the Treaty of Guadalupe-Hidalgo in 1848, claimants to Spanish and Mexican land grants in the ceded territory had to have their claims of ownership con-firmed by a United States Federal Court, after which a patent was issued. The description in the patent was based on a survey of the land in which Gunter's Chain of 66 feet was the unit of measure used. Thus, the vara ceased to have any practical significance in the state of California or other states carved out of the ceded territory.

The Spanish Vara in Texas
Equivalent: 33 $^1/_3$ inches

1 pié	= 1 Spanish foot
1 pié	= 0.92593 English feet
1 vara	= 3 piés
1 vara	= 33 $^1/_3$ inches
1 vara	= 2.777778 feet
1 vara	= 2 feet and 9 $^1/_3$ inches
1 vara	= 0.92593 yards
1 caballería	= 552 × 1,104 varas
1 caballería	= 609,408 square varas
1 caballería	= 107.95 acres (108 acres)
1 labór	= 1,000 varas square
1 labór	= 1,000,000 square varas
1 labór	= 177.136 acres
1 legua	= 1 Spanish league
1 league	= 5,000 varas square
1 league	= 25,000,000 square varas
1 league	= 25 labóres
1 league	= 4,428.4 acres
1 league and labór	= 4,605.5 acres
1 sitio	= 1 square league
1 hacienda	= 5 sitios
1 hacienda	= 22,142 acres
1 acre	= 75.1357173 varas square
1 acre	= 5,645.3760127 square varas
1 mile	= 1,900.8 varas

Comparison of Vara, Yard & Meter
(Comparison of vara, yard and meter is as
1, 1.08 and 1.18 respectively)

1 vara = $33^1/_3$ inches
$^1/_{10}$ vara = $3^1/_3$ inches
$^1/_{100}$ vara = $^1/_3$ inch

1 yard = 3 feet = 36 inches

1 meter = 39.37 inches
$^1/_{10}$ meter = 3.937 inches
$^1/_{100}$ meter = 0.3937 inches = 1 centimeter

Acres to Varas Square and Feet Square

ACRES	VARAS SQUARE	FEET SQUARE
$^1/_4$	37.579	104.355
$^1/_2$	53.129	147.581
1	75.136	208.710
2	106.258	295.161
2 $^1/_2$	118.8	330
3	130.139	361.497
4	150.272	417.421
5	168.009	466.691
6	184.045	511.235
7	198.791	552.196
8	212.516	590.322
9	225.407	626.131
10	237.6	660
20	336.018	933.382
25	375.379	1,043.552
40	475.2	1,320
50	531.289	1,475.804
75	650.694	1,807.484
100	751.357	2,087.103
160	950.4	2,640
200	1,062.579	2,951.609
300	1,301.389	3,614.969
400	1,502.714	4,174.206
500	1,680.086	4,666.905
640	1,900.8	5,280

Varas to Feet

Varas	Feet	Varas	Feet
0.1	0.278	13	36.111
0.2	0.556	14	38.889
0.3	0.833	15	41.667
0.4	1.111	16	44.444
0.5	1.389	17	47.222
0.6	1.667	18	50
0.7	1.944	19	52.778
0.8	2.222	20	55.556
0.9	2.5	21	58.333
1	2.778	22	61.111
2	5.556	23	63.889
3	8.333	24	66.667
4	11.111	25	69.444
5	13.889	26	72.222
6	16.667	27	75
7	19.444	28	77.778
8	22.222	29	80.556
9	25	30	83.333
10	27.778	31	86.111
11	30.556	32	88.889
12	33.333	33	91.667

Varas to Feet—Continued

Varas	Feet	Varas	Feet
34	94.444	55	152.778
35	97.222	56	155.556
36	100	57	158.333
37	102.778	58	161.111
38	105.556	59	163.889
39	108.333	60	166.667
40	111.111	61	169.444
41	113.889	62	172.222
42	116.667	63	175
43	119.444	64	177.778
44	122.222	65	180.556
45	125	66	183.333
46	127.778	67	186.111
47	130.556	68	188.889
48	133.333	69	191.667
49	136.111	70	194.444
50	138.889	71	197.222
51	141.667	72	200
52	144.444	73	202.778
53	147.222	74	205.556
54	150	75	208.333

Varas to Feet—Continued

Varas	Feet	Varas	Feet
76	211.111	97	269.444
77	213.889	98	272.222
78	216.667	99	275
79	219.444	100	277.778
80	222.222	200	555.556
81	225	300	833.333
82	227.778	400	1,111.111
83	230.556	500	1,388.889
84	233.333	600	1,666.667
85	236.111	700	1,944.444
86	238.889	800	2,222.222
87	241.667	900	2,500
88	244.444	1,000	2,777.778
89	247.222	1,100	3,055.556
90	250	1,200	3,333.333
91	252.778	1,300	3,611.111
92	255.556	1,400	3,888.889
93	258.333	1,500	4,166.667
94	261.111	1,600	4,444.444
95	263.889	1,700	4,722.222
96	266.667	1,800	5,000

Varas to Feet—Continued

VARAS	FEET	VARAS	FEET
1,900	5,277.778	3,200	8,888.889
2,000	5,555.556	3,300	9,166.667
2,100	5,833.333	3,400	9,444.444
2,200	6,111.111	3,500	9,722.222
2,300	6,388.889	3,600	10,000
2,400	6,666.667	3,700	10,277.778
2,500	6,944.444	3,800	10,555.556
2,600	7,222.222	3,900	10,833.333
2,700	7,500	4,000	11,111.111
2,800	7,777.778	4,500	12,500
2,900	8,055.556	5,000	13,888.889
3,000	8,333.333		
3,100	8,611.111		

NOTE: 570 varas equals 1,583.33 feet.   In the preceding table of "Varas to Feet," notice that 57 varas equals 158.333 feet. By moving the decimal point one place to the right, the result will be 570 varas equals 1,583.33 feet. If it is desired to convert a combination like 577.8 varas to feet, proceed as follows:

Varas		Feet
500	=	1,388.889
77	=	213.889
0.8	=	2.222
577.8	=	1,605.000

Feet to Varas

Feet	Varas	Feet	Varas
0.1	0.036	13	4.68
0.2	0.072	14	5.04
0.3	0.108	15	5.4
0.4	0.144	16	5.76
0.5	0.18	17	6.12
0.6	0.216	18	6.48
0.7	0.252	19	6.84
0.8	0.288	20	7.2
0.9	0.324	21	7.56
1	0.36	22	7.92
2	0.72	23	8.28
3	1.08	24	8.64
4	1.44	25	9
5	1.8	26	9.36
6	2.16	27	9.72
7	2.52	28	10.08
8	2.88	29	10.44
9	3.24	30	10.8
10	3.6	31	11.16
11	3.96	32	11.52
12	4.32	33	11.88

Feet to Varas—Continued

Feet	Varas	Feet	Varas
34	12.24	55	19.8
35	12.6	56	20.16
36	12.96	57	20.52
37	13.32	58	20.88
38	13.68	59	21.24
39	14.04	60	21.6
40	14.4	61	21.96
41	14.76	62	22.32
42	15.12	63	22.68
43	15.48	64	23.04
44	15.84	65	23.4
45	16.2	66	23.76
46	16.56	67	24.12
47	16.92	68	24.48
48	17.28	69	24.84
49	17.64	70	25.2
50	18	71	25.56
51	18.36	72	25.92
52	18.72	73	26.28
53	19.08	74	26.64
54	19.44	75	27

Feet to Varas—Continued

Feet	Varas	Feet	Varas
76	27.36	97	34.92
77	27.72	98	35.28
78	28.08	99	35.64
79	28.44	100	36
80	28.8	200	72
81	29.16	300	108
82	29.52	400	144
83	29.88	500	180
84	30.24	600	216
85	30.6	700	252
86	30.96	800	288
87	31.32	900	324
88	31.68	1,000	360
89	32.04	1,100	396
90	32.4	1,200	432
91	32.76	1,300	468
92	33.12	1,400	504
93	33.48	1,500	540
94	33.84	1,600	576
95	34.2	1,700	612
96	34.56	1,800	648

Feet to Varas—Continued

Feet	Varas	Feet	Varas
1,900	684	7,500	2,700
2,000	720	8,000	2,880
2,500	900	8,500	3,060
3,000	1,080	9,000	3,240
3,500	1,260	9,500	3,420
4,000	1,440	10,000	3,600
4,500	1,620	10,500	3,780
5,000	1,800	11,000	3,960
5,500	1,980	11,500	4,140
6,000	2,160	12,000	4,320
6,500	2,340	12,500	4,500
7,000	2,520	15,000	5,400

Varas to Yards

Varas	Yards	Varas	Yards
1	0.92593	23	21.29629
2	1.85185	24	22.22222
3	2.77777	25	23.14815
4	3.70370	26	24.07407
5	4.62963	27	24.99999
6	5.55555	28	25.92593
7	6.48148	29	26.85185
8	7.40741	30	27.77777
9	8.33333	31	28.70370
10	9.25926	32	29.62963
11	10.18518	33	30.55555
12	11.11111	34	31.48148
13	12.03704	35	32.40741
14	12.96296	36	33.33333
15	13.88888	37	34.25926
16	14.81481	38	35.18518
17	15.74074	39	36.11111
18	16.66666	40	37.03704
19	17.59259	41	37.96296
20	18.51852	42	38.88888
21	19.44444	43	39.81481
22	20.37037	44	40.47407

Varas to Yards—Continued

Varas	Yards	Varas	Yards
45	41.66666	67	62.03704
46	42.59259	68	62.96296
47	43.51852	69	63.88888
48	44.44444	70	64.81481
49	45.37037	71	65.74074
50	46.29629	72	66.66666
51	47.22222	73	67.59259
52	48.14815	74	68.51852
53	49.07407	75	69.44444
54	49.99999	76	70.37037
55	50.92593	77	71.29629
56	51.85185	78	72.22222
57	52.77777	79	73.14815
58	53.70370	80	74.07407
59	54.62963	81	74.99999
60	55.55555	82	75.92593
61	56.48148	83	76.85185
62	57.40741	84	77.77777
63	58.33333	85	78.70370
64	59.25926	86	79.62963
65	60.18518	87	80.55555
66	61.11111	88	81.48148

Varas to Yards—Continued

Varas	Yards
89	82.40741
90	83.33333
91	84.29526
92	85.18518
93	86.11111
94	87.03704
95	87.96296
96	88.88888
97	89.81481
98	90.74074
99	91.66666
100	92.59259

The Spanish Caballería
Dimensions, Subdivisions and Contents
(Base: 1 vara = 33 $\frac{1}{3}$ inches)

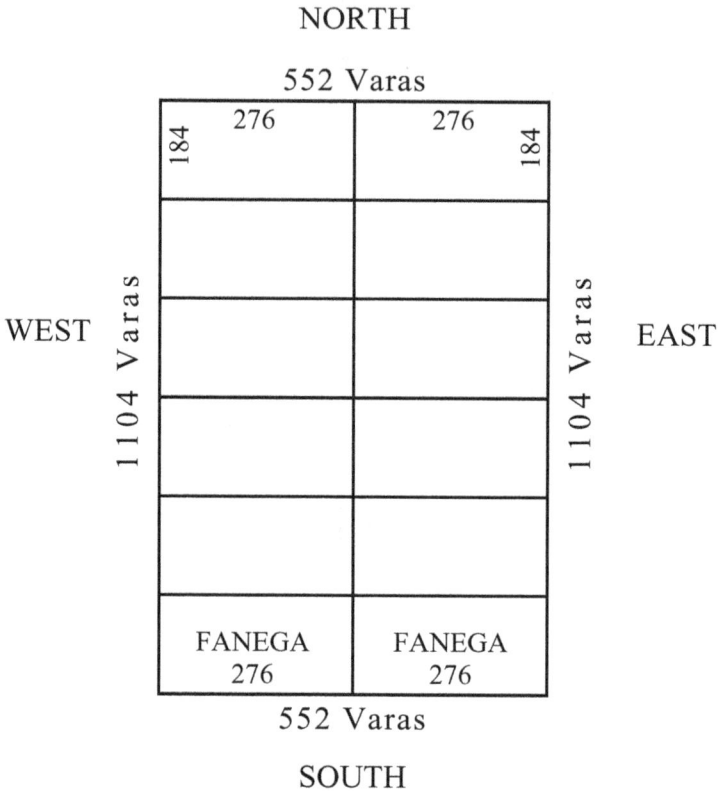

NORTH

552 Varas

184	276	276	184

WEST   1104 Varas   1104 Varas   EAST

	FANEGA 276	FANEGA 276	

552 Varas

SOUTH

The caballería de tierra, or caballería of land, is a right-angled parallelogram measuring 552 x 1104 varas, containing 609,408 square varas, 12 fanegas, 107.948 acres, or 108 acres for practical purposes.

A media caballería, or one half caballería, is 552 varas square, containing 304,704 square varas, 6 fanegas, or 54 acres.

A cuarta caballería or suerte de tierra (one fourth caballería or lot of land), is a right-angled parallelogram measuring 276 x 552 varas, containing 152,352 square varas, 3 fanegas or 27 acres.

A fanega de sembradura de maiz (block of land for planting corn), is a right-angled parallelogram measuring 184 x 276 varas, containing 50,784 square varas, $\frac{1}{12}$ caballería, or 9 acres.

38

Spanish Porción Dimensions

A porción consisting of 1 sitio (25,000,000 square varas or 4,428.4 acres) is approximately 20 cordéles or 1,000 varas wide and 500 cordéles or 25,000 varas (13.2 miles) long.

A porción consisting of 2 sitios (50,000,000 square varas or 8,856.8 acres) is approximately 40 cordéles or 2,000 varas wide and 25,000 varas (13.2 miles) long.

A porción consisting of 2 sitios and 6 caballerías (53,658,448 square varas or 9,504.5 acres) is approximately 50 cordéles or 2,500 varas wide and 430 cordéles or 21,500 varas (11.3 miles) long.

A porción consisting of 2 sitios and 12 caballerías (57,312,896 square varas or 10,152.2 acres) is approximately 50 cordéles or 2,500 varas wide and 460 cordéles or 23,000 varas (12 miles) long.

A sitio de gañado menor, or tract for raising small livestock is $3,333^1/_3$ varas square containing $11,111,111^1/_9$ square varas (18.232 caballerías or 1,968.18 acres).

A sitio de gañado mayor, or tract for raising large livestock, is 100 cordéles (5,000 varas square containing 25,000,000 square varas or 41.023 caballerías or 4,428.4 acres).

A criadero de gañado menor, or breeding farm for small live-stock, is $1,666 \,^2/_3$ varas square containing $2,777,777 \,^1/_3$ square varas ($^1/_4$ sitio de gañado menor, 4.558 caballerías, or 492.045 acres).

A criadero de gañado mayor, or breeding farm for large livestock, is 2,500 varas square containing 6,250,000 square varas ($^1/_4$ sitio de gañado mayor, 10.256 caballerías, or 1,107.1 acres).

French Measures in Quebec

In the Canadian province of Quebec, the French arpent of 180 pieds, or 191.835 feet is still used for all lands originally granted under the Seigniorial Tenure of the French regime. The pied, or French foot, was established in Quebec at 12.789 inches by an Act of Parliament dated July 7, 1919.

French Measures in Michigan & Wisconsin

The arpent was also used in French land grants made along the Detroit River in Michigan and the Fox River in Wisconsin prior to the Treaty of Paris in February 1763. By Act of Congress, section 2, April 25, 1808 supplemental to the Act of March 3, 1807 which regulated grants of land in the territory of Michigan, every person whose claim had been confirmed to a tract of land not exceeding 40 arpents in depth within that part of the territory "to which the Indian title had been extinguished," by virtue of a legal grant made by the French government prior to the Treaty of Paris or a legal grant made by the British government subsequent to that treaty or prior to the treaty of peace between the United States and Great Britain at the close of the Revolutionary War, was entitled to preference in becoming the purchaser of any vacant land adjoining his own on the Detroit River not to exceed 40 arpents in depth behind his tract, nor the quantity of same, etc.

French Measures in Louisiana

The arpent is also still used in parts of Louisiana where it was previously used under both French and Spanish rule. Its customary equivalent as used in Louisiana is 191.835 feet for urban or city property, 191.994 feet for rural or farm lands or 192 feet for general practical purposes.

The earliest legislation enacted in France to regulate grants of land in what was then the Province of Louisiana was the Edict of October 12, 1716. Enacted by King Louis XV, the edict noted that land would be conceded to inhabitants in the proportion of 2 to 4 arpents in width by 40 to 60 arpents in depth. In the secret Treaty of Fontainebleau ratified on November 13, 1762, King Louis XV gifted Louisiana land west of the Mississippi River and New Orleans to his cousin, King Carlos III of Spain. In October 1802, the secret Treaty of San Ildefonso culminated in Spain retroceding Louisiana back to France. The last change in sovereignty came on April 30, 1803 with the signing of the Louisiana Purchase Treaty.

French land measures were used in most land grants made in Louisiana while it was under Spanish rule, since the inhabitants were predominantly French and most of the commands and posts were held by Frenchmen in the Spanish service.

The following excerpts are taken from regulations issued at New Orleans on February 18, 1770:

> Art. 1 - "There shall be granted to each newly arrived family who may wish to establish itself on the borders of the river 6 or 8 arpents in front (according to the means of the cultivator) by 40 arpents in depth, in order that it may have the benefit of cypress wood, which is necessary as it is useful to the inhabitants."

> Art. 4 - "The points formed by the lands on the Mississippi River, leaving in some places but little depth, there may be granted in some cases, 12 arpents in front."

Art. 8 - "No grant in Opelousas, Attakapas and Natchitoches shall exceed one league in front by one league in depth; but when the land granted shall not have that depth, a league and a half in front by a half a league in depth may be granted."

Art. 9 - "To obtain in Opelousas, Attakapas and Natchitoches a grant of 42 arpents in front by 42 arpents in depth, the applicant must make it appear that he is the possessor of 100 head of tame cattle, some horses and sheep, and two slaves to look after them; a proportion of which shall always be observed for the grants to be made in the said places, but none shall ever be made of greater extent than that declared in the preceding article."

During Spanish rule in Louisiana, the colony was divided into Upper and Lower Louisiana, with St. Louis as the seat of the district government and New Orleans as the capitol and seat of the senior government in the province. The senior governor of Louisiana was subordinate to the Spanish Captain-General in Havana, Cuba. The latitude of the town of New Madrid, Missouri (36° 35′ North) was regarded as the dividing line between Upper and Lower Louisiana, making the boundary line approximately 5 3/4 miles north of that portion of the present day Missouri-Arkansas state line, the latitude of which is 36° 30′ North.

With regard to the "league" mentioned in the 1770 regulations issued at New Orleans, the following equivalents are submitted for reference:

1 lieue, or French league, was equal to 84 lineal arpents or 3.05445 miles in then Lower Louisiana.

1/2 lieue, or French league, was equal to 42 lineal arpents or 1.52723 miles in then Lower Louisiana.

1 lieue carrée, or square French league, was equal to 7,056 superficial arpents, or 5,970.985 acres in then Lower Louisiana, or 6,002.5 acres in then Upper Louisiana.

French and Spanish surveyors used 2,500 "toise de Paris" as equivalent to a lineal French league. This gave it a length of 83.333333+ instead of 84 lineal arpents and gave the square league an area of 5,876.584 acres based on one lineal arpent being equal to 191.994 feet.

After the purchase of Louisiana from France by the United States, the U.S. Government Surveyors found the majority of grants of farm lands in the lower portion of the former province to have a depth of 40 arpents, which measured on the ground 116.36 chains or 7,679.76 feet, based on Gunter's 4-pole chain of 66 feet. This gave the lineal arpent a length of 2.909 chains or 191.994 feet exactly, or 192 feet for practical purposes and to facilitate the computation of the area. For ordinary purposes, the arpent de surface (superficial arpent) is equivalent to 0.85 acres. The abbreviation sometimes seen for superficial arpents is simply "P".

French Land Measures and Equivalents Used in the Spanish Province of Upper Louisiana

Linear Measure
(Base: 12 lineal arpents = 35 chains or 2,310 feet)

1 lineal arpent	= 192.5 feet
1 lineal arpent	= 2.9166666+ chains
1 lineal arpent	= 10 lineal perches
1 lineal perche	= 19.25 feet
1 French league (lieue)	= 84 lineal arpents
1 French league (lieue)	= 3.0625 miles
1 French league (lieue)	= 16,170 feet

Superficial Measure
(Base: 288 superficial arpents = 245 acres)

1 superficial arpent	= 0.850694444 acres or 0.8507 acres
1 superficial arpent	= 100 square perches
1 acre	= 1.17551 superficial arpents
1 acre	= 1 superficial arpent and 17.551 square perches
1 square French league (lieue carrée)	= 7,056 superficial arpents
1 square French league (lieue carrée)	= 6,002.5 acres

List of Old Land Measures Used in Louisiana and Quebec With Their English Equivalents

(Listes des anciennes measures françaises de surface agraire employees en Louisiane et dans la Province de Quebec, avec leurs equivalents Anglais)

1 pied Français	=	12 pouces Français
1 French foot	=	12 French inches
1 pied Français	=	12.789 pouces Anglais
1 French foot	=	12.789 English inches
1 pied Anglais	=	12 pouces Anglais
1 English foot	=	12 English inches
1 pied Français	=	1.066 pieds Anglais
1 French foot	=	1.066 English feet
1 pied Anglais	=	0.938 pied Français
1 English foot	=	0.938 French feet
1 verge	=	3 pieds Anglais
1 yard	=	3 English feet
1 chainon	=	7.92 pouces Anglais
1 link	=	7.92 English inches
1 perch	=	25 chainons Anglais
1 perch	=	25 English links
1 perch	=	16 ½ pieds Anglais
1 perch	=	16 ½ English feet

1 perche	= 19.1835 pieds Anglais (Quebec)
1 perche	= 19.1835 English feet (Quebec)
1 perche	= 19.1835 pieds Anglais (Louisiane)
1 perche	= 19.1835 English Feet (Urban Louisiana)
1 perche	= 19.1994 pieds Anglais (Louisiane)
1 perche	= 19.1994 English feet (Rural Louisiana)
1 perche	= 18 pieds Français
1 perche	= 18 French feet
1 toise	= 6 pieds Français
1 toise	= 6 French feet
1 arpent	= 180 pieds Français
1 arpent	= 180 French feet
1 arpent	= 191.835 pieds Anglais (Quebec)
1 arpent	= 191.835 English feet (Quebec)
1 arpent	= 191.835 pieds Anglais (Louisiane)
1 arpent	= 191.835 English feet (Urban Louisiana)
1 arpent	= 191.994 pieds Anglais (Louisiane)
1 arpent	= 191.994 English feet (Rural Louisiana)
1 arpent de surface	= .85 acre
1 acre	= 1.18 superficial arpents

1 chaine	= 66 pieds Anglais
1 chain	= 66 English feet
1 stade	= 10 chaines Anglais
1 furlong	= 10 English chains
1 stade	= 660 pieds Anglais
1 furlong	= 660 English feet
1 mille	= 5,280 pieds Anglais
1 mile	= 5,280 English feet
1 lieue Français	= 84 arpents
1 French league	= 84 arpents
1 ligne	= $\frac{1}{8}$ pouce Anglais (Louisiane)
1 line	= $\frac{1}{8}$ English inch (Louisiana)

The French Arpent
Equivalent: 192 feet

1 lineal arpent	= 192 feet
1 lineal arpent	= 64 yards
1 lineal arpent	= 58.5217 meters
1 lineal arpent	= 11.6364 rods
1 lineal arpent	= 2.9091 chains
1 superficial arpent	= 36,864 square feet
1 superficial arpent	= 0.846281 acres
1 acre	= 1.181641 superficial arpents

The French Arpent
(Quebec and Urban Louisiana)
Equivalent: 191.835 feet, exactly
Base: 1 pied = 12.789 inches

1 ligne (line)	= 1/8 inch (Louisiana)
1 pied de roi	= 1 French foot
1 pied de roi	= 12.789 inches
1 pied de roi	= 1.06575 feet
1 toise	= 6 pieds
1 toise	= 6.3945 feet
1 toise	= 2.13122 yards
1 toise (Louisiana)	= 1 compass
1 perche	= 3 toises
1 perche	= 18 pieds
1 perche	= 19.1835 feet
1 lineal arpent	= 10 perches
1 lineal arpent	= 30 toises
1 lineal arpent	= 180 pieds
1 lineal arpent	= 191.835 feet
1 lineal arpent	= 191 feet, 10 inches
1 lineal arpent	= 69.0606 varas
1 lineal arpent	= 63.945 yards
1 lineal arpent	= 58.471 meters

The French Arpent
(Quebec and Urban Louisiana - Continued)

1 superficial arpent	= 1 lineal arpent squared
1 superficial arpent	= 100 square perches
1 superficial arpent	= 36,800.667 square feet
1 superficial arpent	= 0.845 acres
1 acre	= 1.18369 superficial arpents

The French Arpent
(Rural Louisiana - Farm Lands)
Equivalent: 191.994 feet, exactly

1 pied	= 1.06663 feet
1 toise	= 6.396 feet
1 toise	= 2.132 yards
1 perche	= 19.1994 feet
1 perche	= 6.3998 yards
1 perche	= 1.1636 rods
1 perche	= 5.85198 meters
1 lineal arpent	= 191.994 feet, exactly
1 lineal arpent	= 63.998 yards
1 lineal arpent	= 58.198 meters
1 lineal arpent	= 11.636 rods
1 lineal arpent	= 2.909 chains, exactly
1 superficial arpent	= 36,861.696 square feet
1 superficial arpent	= 0.8462281 acres, exactly
1 acre	= 1.181715 superficial arpents

Comparison of Acre & Superficial Arpent

(Rural Louisiana - Farm Lands)

Base: 1 lineal arpent = 191.994 feet, exactly

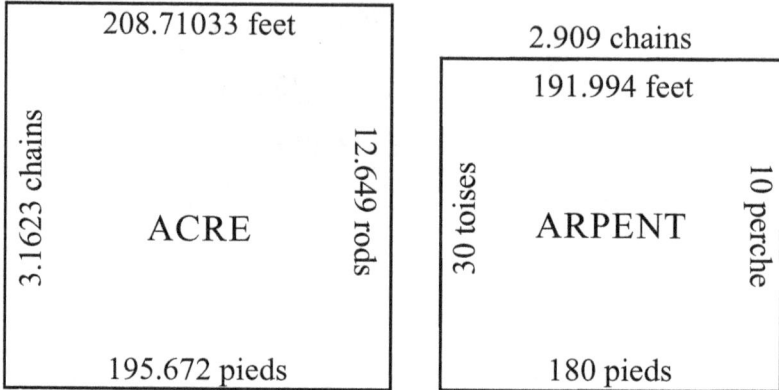

	ACRE	
3.1623 chains	208.71033 feet	12.649 rods
	195.672 pieds	

	ARPENT	
	2.909 chains	
30 toises	191.994 feet	10 perche
	180 pieds	

1 acre	= 43,560 square feet
1 superficial arpent	= 36,861.696 square feet
1 acre	= 1.18171448 superficial arpents
1 superficial arpent	= 0.8462281 acres exactly
1 pied	= 12.7996 inches, exactly
1 pied	= 1.066633334 feet
1 foot	= 0.9375292978 pieds

Lineal Arpents to Feet

Base: 1 lineal arpent = 191.994 feet, exactly

Arpents	Feet	Arpents	Feet
1	191.994	21	4,031.874
2	383.988	22	4,223.868
3	575.982	23	4,415.862
4	767.976	24	4,607.856
5	959.970	25	4,799.850
6	1,151.964	26	4,991.844
7	1,343.958	27	5,183.838
8	1,535.952	28	5,375.832
9	1,727.946	29	5,567.826
10	1,919.940	30	5,759.820
11	2,111.934	31	5,951.814
12	2,303.928	32	6,143.808
13	2,495.922	33	6,335.802
14	2,687.916	34	6,527.796
15	2,879.910	35	6,719.790
16	3,071.904	36	6,911.784
17	3,263.898	37	7,103.778
18	3,455.892	38	7,295.772
19	3,647.886	39	7,487.766
20	3,839.880	40	7,679.760

Superficial Arpents to Acres
Base: 1 lineal arpent = 191.994 feet, exactly
Equivalent: 1 superficial arpent = 0.8462281 acres, exactly

Arpents	Acres	Arpents	Acres
1	0.8462281	22	18.6170182
2	1.6924562	23	19.4632463
3	2.5386843	24	20.3094744
4	3.3849124	25	21.1557025
5	4.2311405	26	22.0019306
6	5.0773686	27	22.8481587
7	5.9235967	28	23.6943868
8	6.7698248	29	24.5406149
9	7.6160529	30	25.3868430
10	8.4622810	31	26.2330711
11	9.3085091	32	27.0792992
12	10.1547372	33	27.9255273
13	11.0009653	34	28.7717554
14	11.8471934	35	29.6179835
15	12.6934215	36	30.4642116
16	13.5396496	37	31.3104397
17	14.3858777	38	32.1566678
18	15.2321058	39	33.0028959
19	16.0783339	40	33.8491240
20	16.9245620	41	34.6953521
21	17.7707901	42	35.5415802

Superficial Arpents to Acres—Continued

Arpents	Acres	Arpents	Acres
43	36.3878083	64	54.1585984
44	37.2340364	65	55.0048265
45	38.0802645	66	55.8510546
46	38.9264926	67	56.6972827
47	39.7727207	68	57.5435108
48	40.6189488	69	58.3897389
49	41.4651769	70	59.2359670
50	42.3114050	71	60.0821951
51	43.1576331	72	60.9284232
52	44.0038612	73	61.7746513
53	44.8500893	74	62.6208794
54	45.6963174	75	63.4671075
55	46.5425455	76	64.3133356
56	47.3887736	77	65.1595637
57	48.2350017	78	66.0057918
58	49.0812298	79	66.8520199
59	49.9274579	80	67.6982480
60	50.7736860	81	68.5444761
61	51.6199141	82	69.3907042
62	52.4661422	83	70.2369323
63	53.3123703	84	71.0831604

Superficial Arpents to Acres—Continued

Arpents	Acres	Arpents	Acres
85	71.9293885	98	82.9303538
86	72.7756166	99	83.7765819
87	73.6218447	100	84.6228100
88	74.4680728	200	169.24562
89	75.3143009	300	253.86843
90	76.1605290	400	338.49124
91	77.0067571	500	423.11405
92	77.8529852	600	507.73686
93	78.6992133	700	592.35967
94	79.5454414	800	676.98248
95	80.3916695	900	761.60529
96	81.2378976	1,000	846.22810
97	82.0841257		

NOTE: In the preceding tables of equivalents, the decimal point may be moved to the right to extend the values. For example, to find the equivalent of 950 arpents in acres, look at 95 arpents above, which are equivalent to 80.3916695 acres. By moving the decimal point one place to the right, 950 arpents equals 803.916695 acres.

Alphabetical List of Land Measures, Quantities and Comparison of Units—Linear and Superficial

- A -

1 acre	= 1.18 superficial arpents
1 acre	= 3.1623 chains square
1 acre	= 10 square chains
1 acre	= 208.71033 feet square
1 acre	= 208 feet, 8 $\frac{1}{2}$ inches square
1 acre	= 43,560 square feet
1 acre	= 0.404687 hectare
1 acre	= 63.616 meters square
1 acre	= 4,046.87 square meters
1 acre	= 4 roods
1 acre	= 12.649 rods square
1 acre	= 160 square rods
1 acre	= 75.1357 varas square
1 acre	= 5,645.376 square varas
1 acre	= 69.57 yards square
1 acre	= 4,840 square yards
1 lineal arpent	= 10 perches
1 lineal arpent	= 180 pieds de roi
1 lineal arpent	= 30 toises

- A -
Continued

1 superficial arpent	= 0.85 acres
1 superficial arpent	= 1 lineal arpents square
1 superficial arpent	= 180 pieds square
1 superficial arpent	= 100 square perches
1 superficial arpent	= 32,400 square pieds
1 superficial arpent	= 30 toises square
1 are	= 100 square meters

- C -

1 caballería	= 108 acres
1 caballería	= 552 × 1,104 varas
1 caballería	= 609,408 square varas
1 caballería	= 12 fanegas de tierra
½ caballería	= 54 acres
½ caballería	= 552 varas square
½ caballería	= 304,704 square varas
½ caballería	= 6 fanegas
¼ caballería	= 27 acres
¼ caballería	= 552 × 276 varas
¼ caballería	= 152,352 square varas
¼ caballería	= 3 fanegas
¼ caballería	= 1 suerte (lot)
1 centare	= 1 square meter
1 centimeter	= 0.3937 inches
1 chain (Engineer's)	= 100 links
1 chain (Engineer's)	= 100 feet
1 chain (Engineer's)	= 36 varas
1 chain (Engineer's)	= 33.333 yards
1 chain (Engineer's)	= 30.48 meters
1 chain (Gunter's)	= 100 links
1 chain (Gunter's)	= 66 feet
1 chain (Gunter's)	= 20.117 meters
1 chain (Gunter's)	= 4 rods

- C -
Continued

1 chain (Gunter's) = 23.76 varas

1 chain (Gunter's) = 22 yards

1 square chain* = 4,356 square feet

1 square chain* = 404.687 square meters

1 square chain* = 16 square rods

1 square chain* = 564.5376 square varas

1 square chain* = 484 square yards

1 square chain* = 0.1 acres

1 square chain* = 10,000 square links

1 compass = 1 toise

* square Gunter's chain

- F -

1 fanega de tierra	= 276 × 184 varas
1 fanega de tierra	= 50,784 square varas
1 fanega de tierra	= $1/12$ caballería
1 fanega de tierra	= 9 acres
1 foot	= 30.48 centimeters
1 foot	= 0.01515 chains
1 foot	= 12 inches
1 foot	= 0.3048 meters
1 foot	= 0.0606 rods
1 foot	= 0.36 varas
1 foot	= 0.3333 ($1/3$) yards
1 square foot	= 144 square inches
1 square foot	= 0.00023 square chains
1 square foot	= 0.0929 square meters
1 square foot	= 0.00367 square rods
1 square foot	= 0.1296 square varas
1 square foot	= 0.1111 ($1/9$) square yards
1 square foot	= 0.000023 acres
1 furlong	= 660 feet
1 furlong	= 220 yards
1 furlong	= 40 rods
1 furlong	= $1/8$ mile
1 furlong	= 201.168 meters

- H -

1 hacienda	= 5 sitios
1 hacienda	= 5 square leagues
1 hacienda	= 22,142 acres
1 hectare	= 24.71 square chains
1 hectare	= 107,638.7 square feet
1 hectare	= 10,000 square meters
1 hectare	= 395.367 square rods
1 hectare	= 13,949.78 square varas
1 hectare	= 11,959.85 square yards
1 hectare	= 2.471 acres
1 hectare	= 4.971 chains square
1 hectare	= 328.083 feet square
1 hectare	= 100 meters square
1 hectare	= 19.884 rods square
1 hectare	= 118.11 varas square
1 hectare	= 109.36 yards square

- I -

1 inch	= 1,000 mils
1 inch	= 25.4 millimeters
1 inch	= 2.54 centimeters
1 inch	= $1/12$ foot
1 inch	= 0.083333 foot
1 circular inch	= circle, 1 inch in diameter
1 circular inch	= 0.7854 square inches
1 square inch	= 1.27324 circular inches

- K -

1 kilometer	= 3,280.833 feet
1 kilometer	= 1,093.6 yards
1 kilometer	= 198.84 rods
1 kilometer	= 1,000 meters
1 kilometer	= 0.62137 miles

- L -

1 labór	= 177.136 acres
1 labór	= 1,000 varas square
1 labór	= 1,000,000 square varas
1 labór	= 0.526 miles square
1 labór	= 2,777.778 feet square
$^1/_2$ labór	= 88.568 acres
$^1/_2$ labór	= 1,388.889 x 2,777.778 feet
$^1/_2$ labór	= 500 x 1,000 varas
$^1/_2$ labór	= 500,000 square varas
$^1/_4$ labór	= 44.284 acres
$^1/_4$ labór	= 500 varas square
$^1/_4$ labór	= 250 x 1,000 varas
$^1/_4$ labór	= 250,000 square varas
$^1/_4$ labór	= 0.263 miles square
1 league	= 4,428.4 acres
1 league	= 25,000,000 square varas
1 league	= 5,000 varas square
1 league	= 25 labóres
1 league	= 6.919 square miles
1 league	= 210.4377 chains square
1 league	= 1 sitio
1 league	= 2.63 miles square

- L -
Continued

$^1/_2$ league	= 2,214.2 acres
$^1/_2$ league	= 3,535.5 varas square
$^1/_2$ league	= 12,500,000 square varas
$^1/_2$ league	= 1.86 miles square
$^1/_3$ league	= 1,476.13 acres
$^1/_3$ league	= 2,886.64 varas square
$^1/_3$ league	= 8,333,333 $^1/_3$ square varas
$^1/_3$ league	= 1.52 miles square
$^1/_4$ league	= 1,107.1 acres
$^1/_4$ league	= 2,500 varas square
$^1/_4$ league	= 6,250,000 square varas
$^1/_4$ league	= 1.32 miles square
1 league & 1 labór	= 4,605.5 acres
1 league & 1 labór	= 5,099 varas square
1 league & 1 labór	= 26,000,000 square varas
1 league & 1 labór	= 26 labóres
1 league & 1 labór	= 2.68 miles square
1 link, Gunter's	= 0.01 chains
1 link, Gunter's	= 0.66 feet
1 link, Gunter's	= 7.92 inches
1 link, Gunter's	= 0.20117 meters

- L -
Continued

1 link, Gunter's	= 0.04 rods
1 link, Gunter's	= 0.2376 varas
1 link, Gunter's	= 0.22 yards
1 square link, Gunter's	= 0.0001 square chains
1 square link, Gunter's	= 0.4356 square feet
1 square link, Gunter's	= 0.0404687 square meters
1 square link, Gunter's	= 0.056454 square varas
1 square link, Gunter's	= 0.0484 square yards

- M -

1 meter	= 0.04971 chains
1 meter	= 3.28083 feet
1 meter	= 39.37 inches
1 meter	= 0.1988 rods
1 meter	= 1.1811 varas
1 meter	= 1.0936 yards
1 square meter	= 1 centare
1 square meter	= 0.00247 square chains
1 square meter	= 10.7639 square feet
1 square meter	= 0.0395 square rods
1 square meter	= 1.3948 square varas
1 square meter	= 1.196 square yards
1 square meter	= 0.000247 acres
1 nautical mile	= 6,080.27 feet
1 nautical mile	= 1,853.25 meters
1 nautical mile	= 1.1516 statute miles
1 nautical mile	= 1 geographical mile
1 square mile	= 640 acres
1 statute mile	= 8,000 links
1 statute mile	= 5,280 feet
1 statute mile	= 1,900.8 varas
1 statute mile	= 1,760 yards

- M -
Continued

1 statute mile	=	1,609.347 meters
1 statute mile	=	880 fathoms
1 statute mile	=	320 rods
1 statute mile	=	80 chains
1 statute mile	=	8 furlongs
1 statute mile	=	1.6094 kilometers
1 statute mile	=	0.8684 nautical miles
1 statute mile	=	27.5 lineal arpents
1/2 statute mile	=	4,000 links
1/2 statute mile	=	2,640 feet
1/2 statute mile	=	950.4 varas
1/2 statute mile	=	880 yards
1/2 statute mile	=	804.7 meters
1/2 statute mile	=	440 fathoms
1/2 statute mile	=	160 rods
1/2 statute mile	=	40 chains
1/2 statute mile	=	4 furlongs
1/2 statute mile	=	0.8047 kilometers
1/2 statute mile	=	0.4342 nautical miles
1/4 statute mile	=	2,000 links
1/4 statute mile	=	1,320 feet
1/4 statute mile	=	475.2 varas

- M -
Continued

$^1/_4$ statute mile	= 440 yards
$^1/_4$ statute mile	= 402.35 meters
$^1/_4$ statute mile	= 220 fathoms
$^1/_4$ statute mile	= 80 rods
$^1/_4$ statute mile	= 20 chains
$^1/_4$ statute mile	= 2 furlongs
$^1/_4$ statute mile	= 0.40235 kilometers
$^1/_4$ statute mile	= 0.2171 nautical miles
1 mil	= 0.001 inch
1 mil	= 0.0254 millimeters
1 mil	= 0.00254 centimeters
1 millimeter	= 0.001 meters
1 millimeter	= 0.1 centimeters
1 millimeter	= 0.03937 inches
1 millonada	= 1,000,000 square varas

- P -

1 perche	= 18 pieds de roi
1 perche	= 19.1835 feet (urban Louisiana)
1 perche	= 19.1835 feet (Quebec)
1 perche	= 19.1994 feet (rural Louisiana)
1 perche	= 3 toises
1 perch (linear)	= 16.5 feet
1 perch (surface)	= 1 square rod
1 pied de roi	= 12.789 inches
1 pied de roi	= 1.066 feet
1 pole (see also *rod*)	= 16.5 feet

- R -

1 rod or pole	= 16.5 feet
1 rod or pole	= 0.25 chains
1 rod or pole	= 25 links
1 rod or pole	= 5.03 meters
1 rod or pole	= 5.94 varas
1 rod or pole	= 5.5 yards
1 square rod	= 0.0625 square chains
1 square rod	= 272.25 square feet
1 square rod	= 25.293 square meters
1 square rod	= 35.284 square varas
1 square rod	= 30.25 square yards
1 square rod	= 625 square links
1 square rod	= 0.00625 acres
1 rood	= $1/4$ acre
1 rood	= 40 square rods
1 rood	= 10,890 square feet

- S -

1 section	= 640 acres
1 section	= 259 hectares
1 section	= 1 square mile
1 section	= 1,900.8 varas square
1 section	= 1,609.347 meters square
1 section	= 5,280 feet square
1 section	= 1,760 yards square
1 section	= 320 rods square
1 section	= 80 chains square
1 sitio	= 1 square league
1 sitio	= 4,428.4 acres
1 sitio	= $1/5$ hacienda
1 sitio de gañado mayor	= 1 square league
1 sitio de gañado mayor	= 4,428.4 acres
1 sitio de gañado mayor	= 5,000 varas square
1 sitio de gañado menor	= 1,968.18 acres
1 sitio de gañado menor	= 3,333 $1/3$ varas square
1 sitio de gañado menor	= 11,111,111 $1/9$ square varas

- T -

1 toise	= 6 pieds de roi
1 toise	= 1 compass
1 toise	= 6.4 feet
1 township	= 6 miles square
1 township	= 36 square miles
1 township	= 36 sections

- V -

1 vara	= 0.04209 chains
1 vara	= 2.77777778 feet
1 vara	= 33 $\frac{1}{3}$ inches
1 vara	= 4.20875 links
1 vara	= 0.8466 meters
1 vara	= 0.1684 rods
1 vara	= 0.92593 yards
$\frac{1}{2}$ vara	= 0.02105 chains
$\frac{1}{2}$ vara	= 1.3889 feet
$\frac{1}{2}$ vara	= 16 $\frac{2}{3}$ inches
$\frac{1}{2}$ vara	= 2.10438 links
$\frac{1}{2}$ vara	= 0.4233 meters
$\frac{1}{2}$ vara	= 0.0842 rods or poles
$\frac{1}{2}$ vara	= 0.46295 yards
1 square vara	= 0.00177136 square chains
1 square vara	= 7.716049 square feet
1 square vara	= 0.7168 square meters
1 square vara	= 0.02834 square rods or poles
1 square vara	= 0.8573 square yards
1 square vara	= 0.000177136 acres

- Y -

1 yard	= 0.04545 chains
1 yard	= 3 feet
1 yard	= 36 inches
1 yard	= 4.545 links
1 yard	= 0.9144 meters
1 yard	= 1.08 varas
1 square yard	= 0.002066 square chains
1 square yard	= 9 square feet
1 square yard	= 0.8361 square meters
1 square yard	= 0.0331 square rods
1 square yard	= 1.1664 square varas
1 square yard	= 0.0002066 acres

Reduction and Conversion Factors

- A -

Reduce/Convert:	To:	Multiply By:
acres	square chains	10
acres	square feet	43,560
acres	square meters	4,046.9
acres	square rods	160
acres	square varas	5,645.376
acres	square yards	4,840
acres	hectares	0.40469
acres	square kilometers	0.00405
acres	superficial arpents (where lineal arpent = 191.835 feet)	1.18369
acres	superficial arpents (where lineal arpent = 191.994 feet)	1.181715
acres	superficial arpents (where lineal arpent = 192 feet)	1.181641
lineal acres	lineal arpents	1.0871 (Louisiana)
superficial arpents	acres (where lineal arpent = 191.835 feet)	0.844827
superficial arpents	acres (where lineal arpent = 191.994 feet)	0.846228
superficial arpents	acres (where lineal arpent = 192 feet)	0.84628
lineal arpents	lineal acres	0.91991 (Louisiana)

- C -

Reduce/ Convert:	To:	Multiply By:	or	Divide By:
chains	feet	66		
chains	meters	20.117		
chains	rods or poles	4		
chains	varas	23.76		
chains	yards	22		
square chains	acres	0.1		10
square chains	square feet	4,356		
square chains	square meters	404.69		
square chains	square rods	16		
square chains	square varas	564.54		
square chains	square yards	484		0.002066

- F -

Reduce/ Convert:	To:	Multiply By:	or	Divide By:
feet	chains	0.1515		66
feet	meters	0.3048006		3.2808
feet	rods	0.0606		16.5
feet	varas	0.36		
feet	yards			3
lineal feet	miles	19 and point off five decimal places (0.00019)		5,280
decimals of a foot	inches	12		
square feet	acres			43,560
square feet	acres	23 and point off six decimal places (0.000023)		
square feet	acres	22,957 and point off nine decimal places (0.000022957)		
square feet	square chains	0.00022957		4,356
square feet	square meters	0.0929		
square feet	square rods or square poles	0.00367		
square feet	square varas	0.1296		
square feet	square yards			9

- H -

Reduce/ Convert:	To:	Multiply By:	or	Divide By:
hectares	acres	2.471		

- I -

Reduce/ Convert:	To:	Multiply By:	or	Divide By:
inches	millimeters	25.4		
inches	decimals of a foot	0.08333		12

- K -

Reduce/ Convert:	To:	Multiply By:	or	Divide By:
kilometers	miles	0.621377		
square kilometers	acres	247.1098		

- M -

Reduce/ Convert:	To:	Multiply By: or	Divide By:
meters	chains	0.04971	
meters	feet	3.2808	
meters	rods	0.1988	
meters	varas	1.1811	
meters	yards	1.0936	
lineal meters	miles	6,214 and point off seven decimal places (0.0006214)	1,609.4
square meters	acres	0.0002471	4,046.9
square meters	square chains	0.002471	
square meters	square feet	10.7639	
square meters	square rods	0.0395	
square meters	square varas	1.13949	
square meters	square yards	1.196	
square meters	hectares	0.0001 (or simply point off 4 decimal places)	
miles	kilometers	1.6094	
millimeters	inches	0.03937	

- R -

Reduce/ Convert:	To:	Multiply By: or	Divide By:
rods	chains	0.25	4
rods	feet	16.5	
rods	meters	5.03	
rods	varas	5.94	
rods	yards	5.5	
square rods or poles	acres	625 & point off five decimal places (0.00625)	160
square rods or poles	square chains	0.0625	16
square rods or poles	square feet	272.25	
square rods or poles	square meters	25.293	
square rods or poles	square varas	35.284	
square rods or poles	square yards	30.25	
lineal rods or poles	miles	3,125 & point off six decimal places (0.003125)	.320

- V -

Reduce/ Convert:	To:	Multiply By:	or	Divide By:
varas	chains	0.04209		
varas	feet	2.7778		0.36
varas	meters	0.8466		
varas	rods or poles	0.1684		5.94

TIP: A convenient method for converting varas to rods/poles is to add one vara for each 100 varas or fraction over 50, then divide by 6. For example, 478 varas plus 5 equals 483, then divided by 6 equals 80.5 rods or poles.

Reduce/ Convert:	To:	Multiply By:	or	Divide By:
varas	yards			1.08
square varas	square chains	0.00177136		
square varas	square feet	7.716049		
square varas	square meters	0.7168		
square varas	square rods	0.02834		
square varas	square yards	0.8575		

To reduce or convert square varas to acres, multiply by 177 and point off six decimal places (0.000177), or divide by 5645.4. Either will give a result correct to within $8/10,000$ of each acre, based on 1 acre containing 5,645.376 square varas. On a calculation of as much as 100 acres, the result will therefore be correct to within $8/100$ of 1 acre.

For more precise conversion of square varas to acres, multiply by 177,136 and point off nine decimal places (0.000177136), or divide by 5,645.376. Either will give a result correct to within $7/10,000,000$ of each acre, based on 1 acre containing 5,645.376 square varas. On a calculation of as much as 100 acres, the result will therefore be correct to within $7/100,000$ of 1 acre.

- V- Continued

Reduce/ Convert:	To:	Multiply By:	or	Divide By:
lineal varas	miles	526 and point off 6 decimal places (0.000526)		
lineal varas	miles	53 and point off 5 decimal places (0.00053)		1,900.8

- Y -

Reduce/ Convert:	To:	Multiply By: or	Divide By:
yards	chains	0.0455	22
yards	feet	3	
yards	meters	0.9144	
yards	rods	0.1818	
yards	varas	1.08	
square yards	acres	2,066 and point off seven decimal places (0.0002066)	4,840
square yards	square chains	0.002066	484
square yards	square feet	9	
square yards	square meters	0.8361	
square yards	square rods or poles	0.0331	30.25
square yards	square varas	1.1664	
lineal yards	miles	57 and point off five decimal places (0.00057)	1,760

The Surveyor's Chain

The surveyor's chain, or Gunter's chain, is 66 feet long and was invented in 1620 by Edmund Gunter (1581–1626), an English mathematician and engineer. It has 100 links, each of which is 7.92 inches long and is based on the fact that 43,560 square feet equal 1 acre and 10 square chains of 66 feet equal 43,560 square feet.

$$(66 \times 66 \times 10 = 43{,}560 \text{ square feet} = 1 \text{ acre})$$

The surveyor's chain was especially valuable in surveying vast areas, such as the public domain of the United States. It also simplifies the computation of area, making the determination of acreage only a matter of ascertaining the number of square chains, then moving the decimal point one place to the left.

The Engineer's Chain

The engineer's chain is 100 feet (36 varas) in length and has 100 links, each of which is 1 foot in length.

Steel Tapes

Flat steel tapes $1/8$ to $1/4$ inch wide and usually 50 or 100 feet in length, graduated to feet and decimals of a foot, are used extensively in modern engineering practice. The steel tape is more convenient, has finer subdivisions and is therefore more accurate than a chain.

A 20-vara tape is 55.556 feet in length. A 40-vara tape is 111.111 feet in length and a 50-vara tape is 138.889 feet in length. Vara tapes have one vara at each end divided decimally. $1/10$ vara equals $3\,1/3$ inches or 0.2778 feet and $1/100$ vara equals $1/3$ inch or 0.02778 feet.

Trigonometric Formulas for Field Note Calculators

Required	Given	Formula
latitude	bearing & distance	distance × cosine of bearing
latitude	bearing & departure	departure × cotangent of bearing
departure	bearing & distance	Distance × sine of bearing
departure	bearing & latitude	latitude × tangent of bearing
sine of bearing	departure & distance	departure ÷ distance
cosine of bearing	latitude & distance	latitude ÷ distance
tangent of bearing	latitude & departure	departure ÷ latitude
cotangent of bearing	latitude & departure	latitude ÷ departure
secant of bearing	distance & latitude	distance ÷ latitude
cosecant of bearing	distance & departure	distance ÷ departure
distance	departure & bearing	departure ÷ sine of bearing
distance	latitude & bearing	latitude ÷ cosine of bearing
distance	latitude & bearing	latitude × secant of bearing
distance	departure & bearing	departure × cosecant of bearing
distance	latitude & departure	$\sqrt{(\text{latitude})^2 + (\text{departure})^2}$

Graphic Solution of Field Notes

(Determination of Area) — *Feet or Varas*

Scale 1 inch equals 100 feet or 36 varas, 1 acre is 2.0871 inches square. 1 square inch equals 0.229568 acres.

Scale 1 inch equals 200 feet or 72 varas, 1 acre is 1.04355 inches square. 1 square inch equals 0.9182736 acres.

Scale 1 inch equals 300 feet or 108 varas, 1 acre is 0.69566 inches square. 1 square inch equals 2.0661157 acres.

Scale 1 inch equals 400 feet or 144 varas, 1 acre is 0.52176 inches square. 1 square inch equals 3.67309 acres.

Scale 1 inch equals 500 feet or 180 varas, 1 acre is 0.41742 inches square. 1 square inch equals 5.7392 acres.

Scale 1 inch equals 600 feet or 216 varas, 1 acre is 0.34785 inches square. 1 square inch equals 8.26446 acres.

Scale 1 inch equals 800 feet or 288 varas, 1 acre is 0.261 inches square. 1 square inch equals 14.692378 acres.

Scale 1 inch equals 1,000 feet or 360 varas, 1 acre is 0.20871 inches square. 1 square inch equals 22.95684 acres.

Graphic Solution of Field Notes
(Determination of Area) — *Varas*

Scale 1 inch equals 75 varas, 1 acre is 1.0018 inches square. 1 square inch equals 0.99639 acres, or for practical purposes 1 acre is 1 inch square and 1/4 acre is 1/2 inch square. 1 mile on plat equals 25.3 inches.

Scale 1 inch equals 100 varas, 1 acre is 0.75136 inches or 3/4 inches square. 1 square inch equals 1.77136 acres. 1 mile on plat equals 19.008 inches.

Scale 1 inch equals 150 varas, 1 acre is 0.5004 inches square. 1 square inch equals 3.9855 acres, or for practical purposes 1 acre is 1/2 inch square, 1/4 acre is 1/4 inch square and 1 inch square equals 4 acres. 1 mile on plat equals 12.68 inches.

Scale 1 inch equals 200 varas, 1 acre is 0.37568 inches or 3/8 inches square. 1 square inch equals 7.085 acres. 1 mile on plat equals 9 1/2 inches.

Scale 1 inch equals 300 varas, 1 acre is 0.25045 inches square. 1 square inch equals 15.942 acres, or for practical purposes 1 acre is 1/4 inch square, 1/4 acre is 1/8 inch square and 1 square inch equals 16 acres. 1 mile on plat equals 6.34 inches.

Scale 1 inch equals 400 varas, 1 acre is 0.18784 inches or 3/16 inches square. 1 square inch equals 28.342 acres. 1 mile on plat equals 4.75 inches.

Scale 1 inch equals 500 varas, 1 acre is 0.15027 inches square. 1 square inch equals 44.284 acres. 1 mile on plat equals 3.8 inches.

Scale 1 inch equals 600 varas, 1 acre is 0.1252 inches or 1/8 inch square. 1 square inch equals 63.769 acres or 64 acres for practical purposes. 1 mile on plat equals 3.17 inches.

Graphic Solution of Field Notes
(Determination of Area) — *Rods or Poles*

Scale 1 inch equals 10 rods, 1 square inch equals 0.625 acres.

Scale 1 inch equals 20 rods, 1 square inch equals 2.5 acres.

Scale 1 inch equals 30 rods, 1 square inch equals 5.625 acres.

Scale 1 inch equals 40 rods (10 chains, 660 feet, 220 yards, 237.6 varas or $1/8$ mile), 1 square inch equals 10 acres. A 2 $1/8$ acre block is $1/2$ inch square. 1 section of 640 acres will be 8 inches square. 1 mile on plat equals 8 inches.

Scale 1 inch equals 50 rods, 1 acre is 0.253 inches square, or approximately $1/4$ inch square. 1 square inch equals 15.625 acres or 15 $5/8$ acres. 1 mile on plat equals 6.4 inches.

Scale 1 inch equals 80 rods (20 chains, 1320 feet, 440 yards, 475.2 varas or $1/4$ mile), 1 square inch equals 40 acres. A 2 $1/2$ acre block is $1/4$ inch square. 1 section of 640 acres will be 4 inches square. 1 mile on plat equals 4 inches.

Scale 1 inch equals 100 rods, 1 square inch equals 62.5 acres.

Length of Unknown Lines

To find the length of an unknown side of a rectangular tract:

If the unit of length is *feet,* multiply the number of acres in the tract by 43,560 (the number of square feet in an acre) and divide the result by the length of the known side.

If the unit of length is *varas,* multiply the number of acres in the tract by 5,645.4 (the number of square varas in an acre) and divide the result by the length of the known side.

If the unit of length is *yards,* multiply the number of acres in the tract by 4,840 (the number of square yards in an acre) and divide the result by the length of the known side.

If the unit of length is *rods or poles,* multiply the number of acres in the tract by 160 (the number of square rods or poles in an acre) and divide the result by the length of the known side.

If the unit of length is *chains,* multiply the number of acres in the tract by 10 (the number of square chains in an acre) and divide the result by the length of the known side.

Constants

π = 3.141592653589793238462643383279502884197169399375 10+

Since the first edition of this work was published in 1946, the computation of π (pi) has grown from 620 digits to well over 6,000,000,000. However, for practical purposes such as the calculation of acreage conversion factors for use with precision instruments, the value of π to 4 or 5 decimal places (π = 3.1416 or π = 3.14159) is sufficient to calculate the equatorial circumference of the earth to within the nearest tenth of an inch.

$\frac{1}{6}\pi$ = 0.52359877559

$\frac{1}{3}\pi$ = 1.04719755119

$\frac{1}{2}\pi$ = 1.57079632679

$\sqrt{2}$ = 1.41421356237

$\frac{\sqrt{2}}{2}$ = 0.70710678119

$\sqrt{3}$ = 1.73205080757

$\frac{\sqrt{3}}{2}$ = 0.86602540378

$\frac{\sqrt{3}}{3}$ = 0.57735026919

π^2 = 9.86960440

$\sqrt{\pi}$ = 1.77245385

$\frac{4\pi}{3}$ = 4.18879020

$\frac{1}{2\pi}$ = 0.15915494

$\frac{1}{\sqrt{\pi}}$ = 0.56418958

$\frac{1}{\pi}$ = 0.31830989

$\frac{4}{\pi}$ = 1.27323954

$\sqrt{\frac{4}{\pi}}$ = 1.12837917

$\frac{1}{4}\pi$	= 0.78539816339744830961	5π	=	15.70796326794896619231
π	= 3.14159265358979323846	6π	=	18.84955592153875943078
2π	= 6.28318530717958647693	7π	=	21.99114857512855266924
3π	= 9.42477796076937971539	8π	=	25.13274122871834590770
4π	= 12.56637061435917295385	9π	=	28.27433388230813914616

Radians, Arcs & Degrees

1 radian = $180° \div \pi$ = 57.296°

arc of 1° = $\pi \div 180$ = 0.01745 radians

degrees x 0.01745 = radians

radians x 57.296 = degrees

Dimensions of a Circle

To Find:	Calculate:
area of a circle	(square of radius) x (3.1416) or πR^2
area of a circle	($1/2$ circumference) x (radius)
area of a circle	(square of diameter) x ($1/4\pi$)
area of a circle	(circumference)² x (0.07958)
area of a circle	(circumference) x ($1/4$ diameter)
area of a circle	(circumference) x ($1/2$ radius)
area of a sector	(arc length) x ($1/2$ radius)
area of a sector	(arc length) x ($1/4$ diameter)
area of a segment	(area of sector) - (area of triangle formed by the radii & chord of the arc of the segment)
length of an arc	(central angle in radians) x (radius)
length of an arc	(circumference x degrees in arc) $\div$ (360)

Dimensions of a Circle—Continued

To Find:	Calculate:
length of an arc	(π x diameter x number of degrees) ÷ (360)
circumference	(radius) x (2π or 6.2832)
circumference	(diameter) x (π or 3.1416)
circumference	($\sqrt{area}$) x (3.5449)
circumference	(diameter) ÷ (0.31831)
circumference circumscribed around a square	(side of square) x (4.44288)
circumference inscribed inside a square	(side of square) x (π)
circumference inscribed inside a square	(circle diameter) x (π)
circumference of equal area as square	(side of square) x (3.545)
diameter 1-acre circle	(= 235.5 feet)
diameter	(circumference x 0.31831)
diameter	($\sqrt{area}$) x (1.1284)
diameter	($\sqrt{area}$) ÷ (0.7854)
diameter	(circumference) ÷ (π)
diameter equal periphery as square	(side of square) x (1.27324)

Dimensions of a Circle—Continued

To Find:	Calculate:
diameter circumscribed around a square	(side of square) x (1.41421)
diameter of equal area as a square	(side of square) x (1.128)
diameter circumscribed around equilateral triangle	(side of triangle) x (1.15185)
diameter inscribed in pentagon	(length of one side) x (1.3764)
diameter circumscribed around pentagon	(length of one side) x (1.7014)
diameter inscribed in hexagon	(length of one side) x (1.732)
diameter circumscribed around hexagon	(length of one side) x (2)
radius	(diameter) x (0.5)
radius	(circumference) x (0.159155)
radius	$(\sqrt{area})$ x (0.56419)
radius	(circumference) ÷ (6.2832)
radius equal periphery as square	(side of square) x (0.63662)
radius circumscribed around square	(side of square) x (0.70711)
radius circumscribed in square	(side of square) ÷ (2)
radius circumscribed around equilateral triangle	(side of triangle) x (0.57592)
radius inscribed in pentagon	(length of one side) x (0.6882)

Dimensions of a Circle—Continued

To Find:	Calculate:
radius circumscribed around pentagon	(length of one side) x (0.8507)
radius inscribed in hexagon	(length of one side) x (0.866)
diameter circumscribed around hexagon	(length of one side)
radius in degrees	(57.296 degrees)
radius in degrees	(57 degrees, 17 minutes & 45 seconds)

Dimensions of a Square

To Find:	Calculate:
area	(square of one side)
diagonal	(side) x (1.41421)
side	(diagonal) x (0.70711)
diagonal inscribed in circle	(diameter of circle)
side	(diagonal) x (0.70711)
side of equal periphery as circle	(diameter of circle) x (0.7854)
side inscribed in circle	(diameter of circle) x (0.70711)
side of square of equal area as circle	(circle diameter) x (0.8862)
side of square of equal area as circle	(circle diameter) ÷ (1.1284)
side of square of equal area as circle	(circle circumference) x (0.2821)
side of square of equal area as circle	(circle circumference) ÷ (3.545)

Scales of Miles for Ownership Maps

SCALE 1 INCH =	1 MILE ON MAP =
1,000 feet	5.28 inches
2,000 feet	2.64 inches
3,000 feet	1.76 inches
4,000 feet	1.32 inches
800 varas	2.375 inches
1,000 varas	1.9 inches
1,500 varas	1.27 inches
2,000 varas	0.95 inch

The Section

5,280 Feet
1,760 Yards

Regular Section

640 acres

1,609.4 Meters   1,900.8 Varas   320 Rods or Poles   80 Chains

1 Mile

1.6094 Kilometers

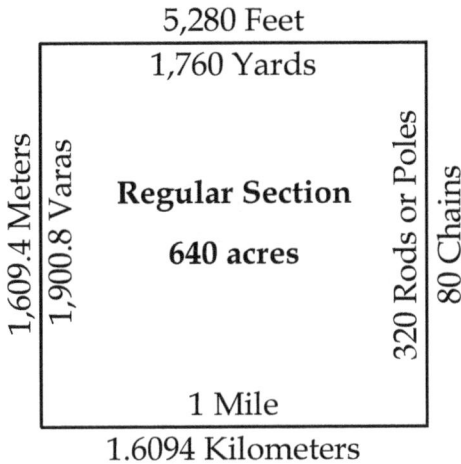

A regular section is 1 mile square and contains 640 acres. An elongated section is an oversized section containing in excess of 640 acres and is usually found along the northern or western boundary of a township or block. It may also be found anywhere within a township or block after fragmentary subdivisions have been made. Elongated sections that border on the boundary line of a township or block will rarely, if ever, be found to exceed 1 1/2 miles in length. Where conditions existed that would permit the designation of elongated sections in excess of 1 1/2 miles in height along a northern boundary, a new half-township or block was usually created.

A fractional section is a section containing less than 640 acres and is usually found along the western line of a township where the boundary cannot be carried out in full due to the convergence of meridians.

Description of a Normal Township

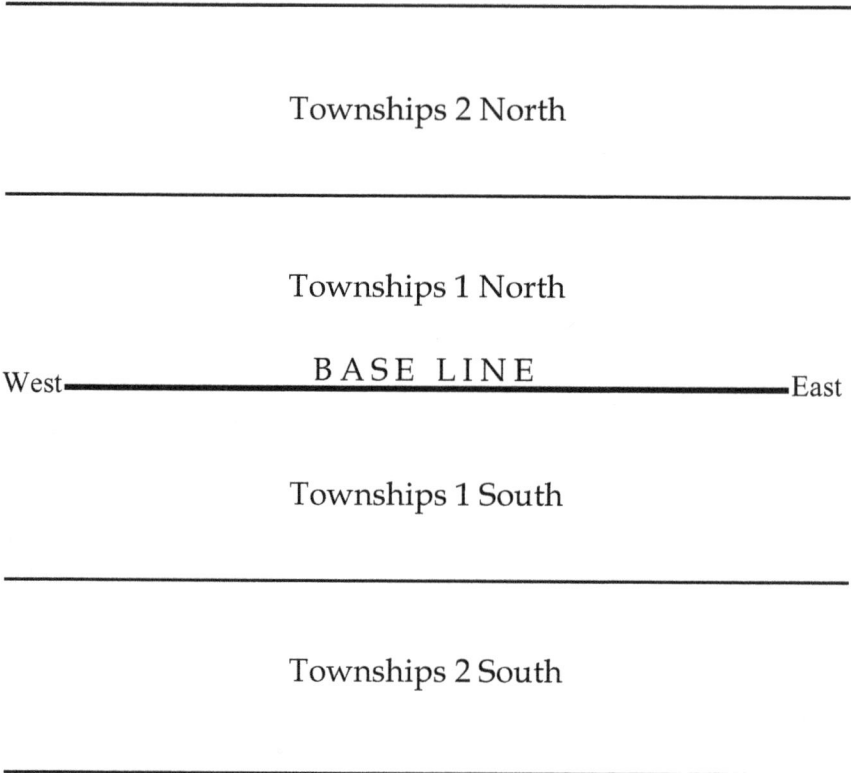

Townships 2 North

Townships 1 North

West ━━━━━━━━━━ **B A S E   L I N E** ━━━━━━━━━━ East

Townships 1 South

Townships 2 South

A normal township is 6 miles square and contains 36 sections.

The rows of townships running east and west are numbered 1, 2, 3, etc. north or south as the case may be, from the base line.

Description of a Range

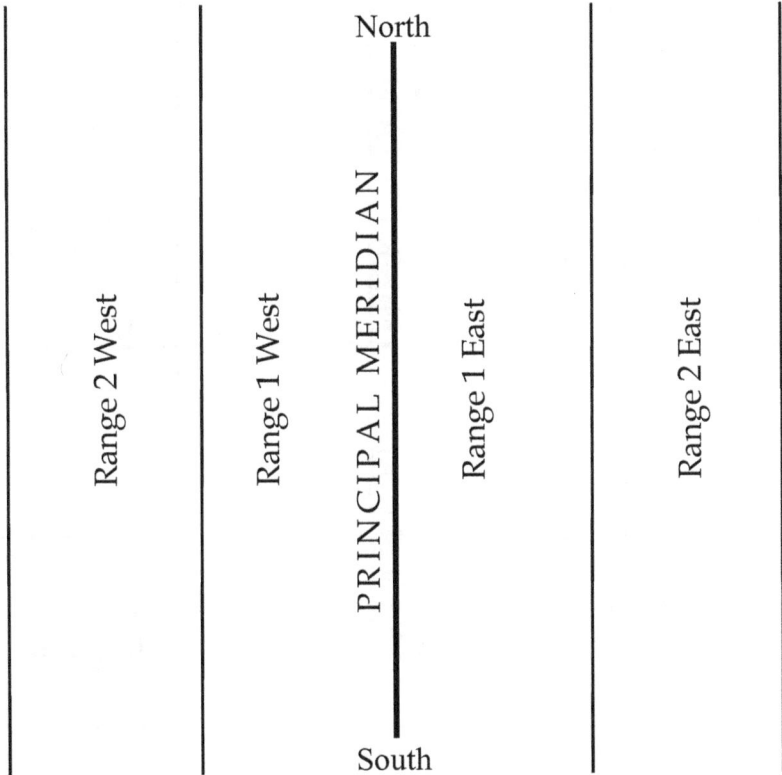

North

PRINCIPAL MERIDIAN

Range 2 West

Range 1 West

Range 1 East

Range 2 East

South

A range is a tier of townships 6 miles wide, located on either side of the principal meridian.

Ranges are numbered 1, 2, 3, etc. east or west, as the case may be, from the principal meridian.

Description of Township & Range Combined

North			
T 2 N R 2 W	T 2 N R 1 W	T 2 N R 1 E	T 2 N R 2 E
T 1 N R 2 W	T 1 N R 1 W	T 1 N R 1 E	T 1 N R 2 E
B A S E		L I N E	
T 1 S R 2 W	T 1 S R 1 W	T 1 S R 1 E	T 1 S R 2 E
T 2 S R 2 W	T 2 S R 1 W	T 2 S R 1 E	T 2 S R 2 E
South			

(vertical text through center: PRINCIPAL MERIDIAN)

The base line is extended both east and west from the initial point on a true parallel of latitude.

The principal meridian is extended both north and south from the initial point on a line conforming to the true meridian of longitude.

The initial points are the points from which the lines of the public land surveys are extended and are established astronomically.

On the plat above, the initial point is shown as the point where the base line and the principal meridian intersect.

Subdivisions of a Normal Township

6	5	4	3	2	1
7	8	9	10	11	12
18	17	16	15	14	13
19	20	21	22	23	24
30	29	28	27	26	25
31	32	33	34	35	36

Enlarged plat of Township 1 South, Range 2 East. Sections in a normal township are numbered as shown on the above plat.

Enlarged plat of Section 26, Township 1 South, Range 2 East.

The 40 acre tract above is an example of which the following is a description:

40 acres, the SW $1/4$ of the SE $1/4$ of Sec. 26, Twp. 1 S, R. 2 E.

Index

Index

Index

www.ingramcontent.com/pod-product-compliance
Lightning Source LLC
Chambersburg PA
CBHW030943150426
42812CB00065B/3152/J